The

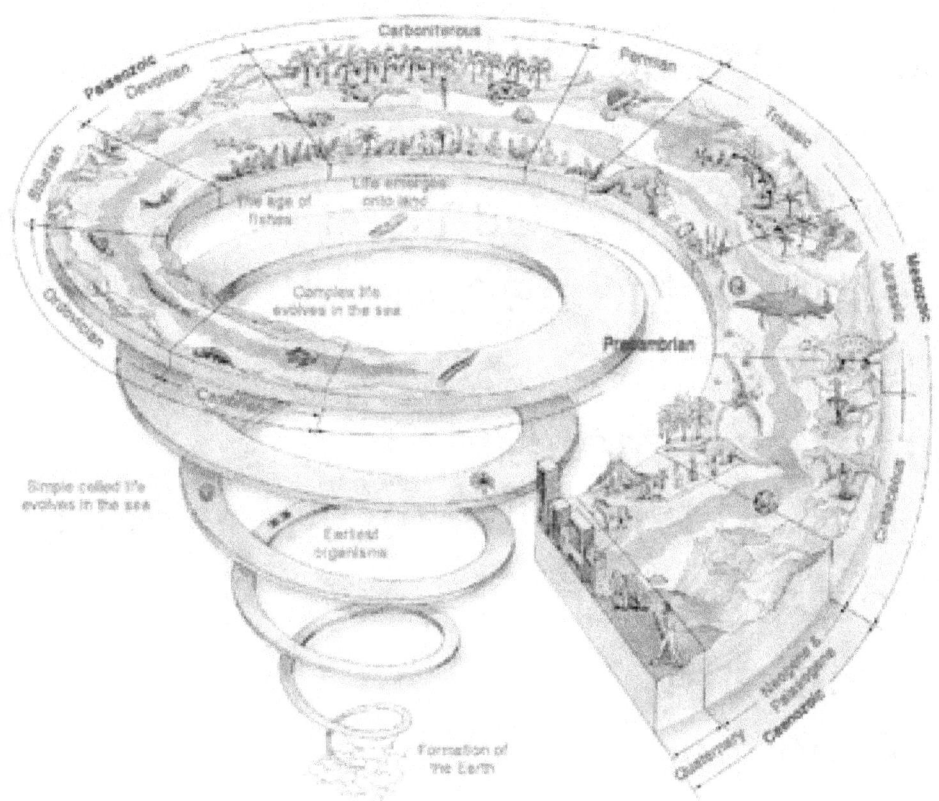

History of the Earth

Monica Sevilla

Contents

Setting the Stage for Life
The Formation of the Universe
The Geologic Time Scale and Life on Earth
The History of the Earth
What is a Fossil?
Precambrian Time: Life Begins
Cyanobacteria: Making Oxygen on Earth
The Proterozoic Eon
Snowball Earth
Mass Extinction

History of the Earth: Setting the Stage for Life

14 billion years ago: The **Big Bang** occurs and our universe was created. Hydrogen atoms form as the universe cools. These atoms consist of 1 proton and 1 electron. Subatomic particles such as protons and electrons were randomly moving around in a plasma before this.

As the universe continues to cool down, **hydrogen atoms** clump together and form stars. Stars group together to form galaxies. Some of these stars explode in a supernova event and send heavier elements and atoms into space.

4.6 billion years ago: The **Solar System** forms from the "left overs" of dying stars. This left over material clumps together into a solar nebula. The sun is created at the center of the solar nebula. The material that is left over produces the planets, the moons, the asteroids, and the comets.

The Formation of the Earth:

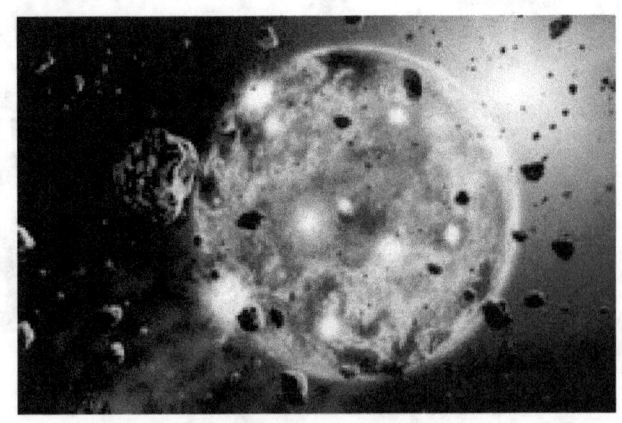

The **Earth** is formed from many heavy elements and atoms such as iron, nickel, and other elements. This material is hot and molten and takes many millions of years to cool down to form the crust. The iron and the nickel sink into the core of the Earth. The lighter materials and elements float above the core to form the mantle and the crust.

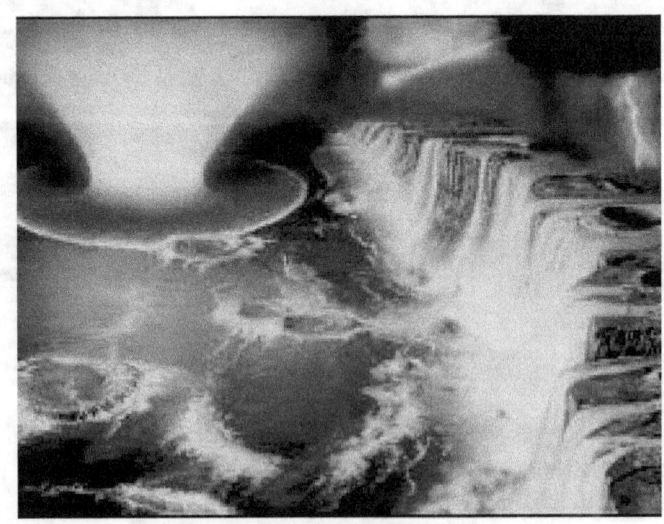

Water starts to accumulate in depressions on the crust of the Earth through asteroid and comet impacts and also from the condensation of steam (water vapor) originating from the molten rock within the mantle inside the Earth. Lakes and oceans developed over a course of millions of years.

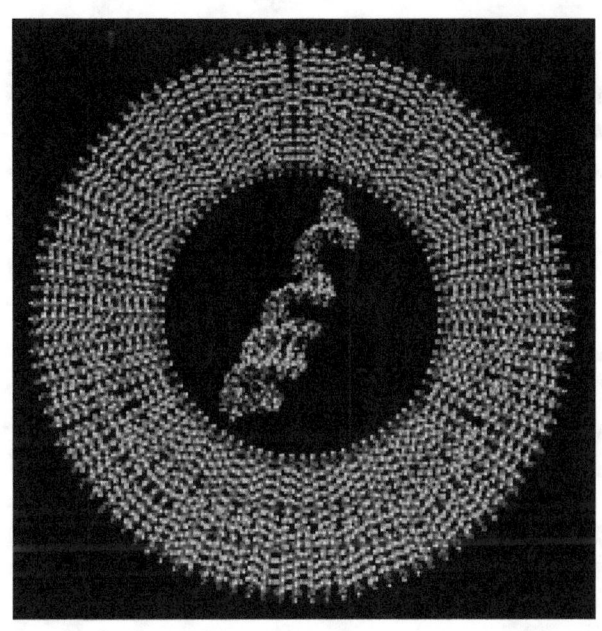

The Earth at this time has no atmosphere and also did not have free oxygen. Inorganic molecules such as carbon dioxide, nitrogen, sulfur dioxide, and water that were present on the surface of the Earth joined together to form the first carbon-based compounds and amino acids. The first "**protocells**" were formed from carbon-based molecules that were present in the water. These protocells eventually evolved into the first organisms, bacteria, about 3.6 billion years ago. These organisms are anaerobic and do not require oxygen to survive.

Cyanobacteria, a photosynthetic bacteria, develops within
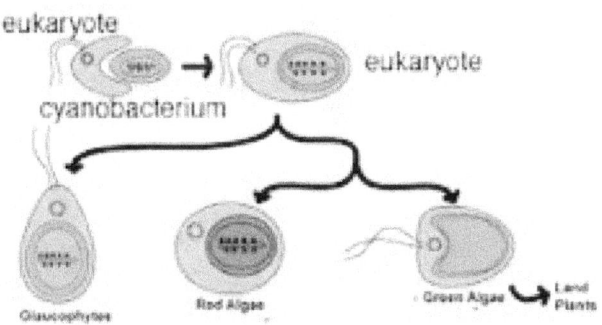
the oceans. It is the first known organism to use carbon dioxide present in the water to form oxygen as a waste product. This oxygen accumulates over millions of years, building up an oxygen-rich atmosphere. This sets the stage for more complex organisms to develop during the

Precambrian Era.

Knowledge and Comprehension:

Big Bang:

Hydrogen atoms:

Solar System:

Earth:

Protocells:

Cyanobacteria:

1. Describe, in one paragraph, how the stage was set for life to have been created on Earth.

2. Identify the major events that occurred:

14 billion years ago

4.6 billion years ago

3.6 billion years ago

3. Which of these events was the most important. Explain why you think so.

Application, Analysis, Evaluation, Synthesis

4. How did cyanobacteria change the Earth?

5. Explain how the first "protocells" formed. Why did they form in the water?

6. Explain how water developed on Earth.

The Formation of the Universe

14.7 billion years ago, when the universe first began, all matter existed within a very small volume. It was very dense and extremely hot. The universe, according to most scientists, was created through the Big Bang. The **Big Bang theory** states that the universe was formed

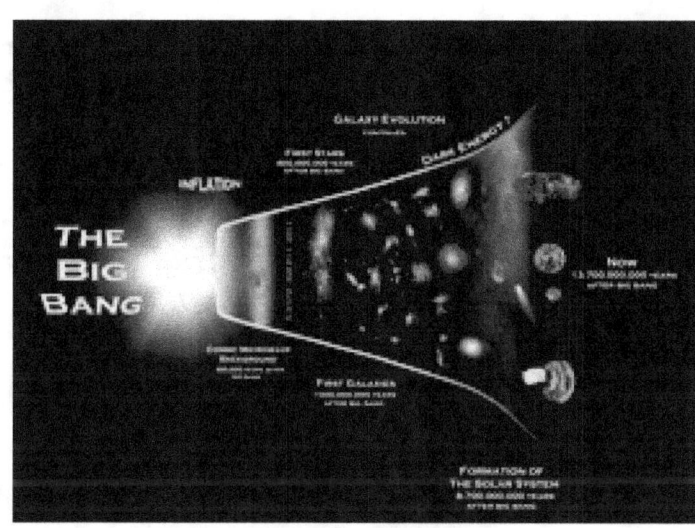

from a single point of **matter** within space and time.

The early universe was extremely hot, but as it cooled, sub atomic particles began to clump together and formed the first atom, the hydrogen atom. Later, as the universe began to cool even further, dust and gas joined and became dense enough to form the first stars, the **protostars**. As hydrogen gas atoms began to fuse together, light and

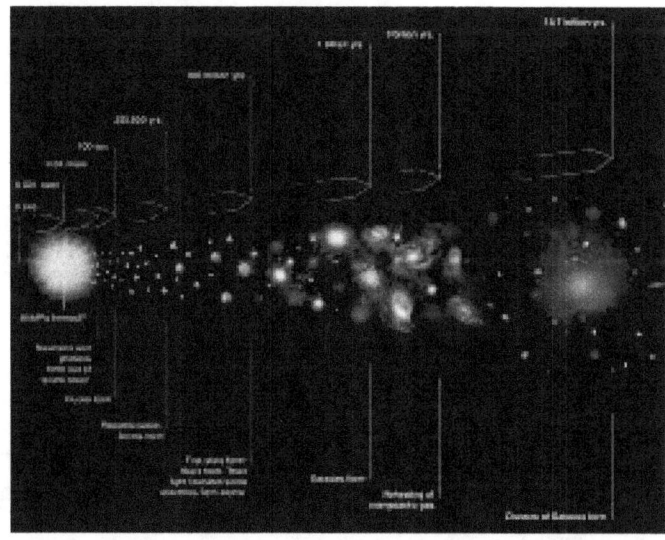

heat energy were produced allowing the **stars** to shine. Billions and billions of stars were created through this process and grouped together to form the **galaxies** within the universe today.

The heavier elements in the periodic table were created within the core of massive stars such as super giant stars. These heavier elements were spread out across the universe after these massive stars ran out of fuel. The collapse and death of these stars caused supernovas. **Supernovas** occur when all of the matter and gas within the outer layers of these stars were forced out through space by a massive galactic explosion. This matter was eventually used to form planets and other bodies within the universe.

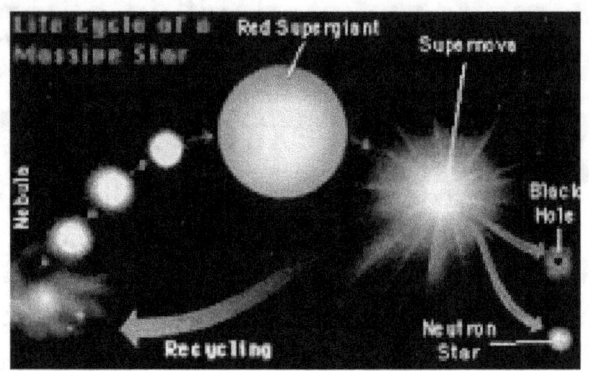

The **Solar System** was formed by the collision, clumping, and cooling of this heavier matter. As this matter progressively clumped together and gained more mass, the force of gravity took over, attracting more material, at a faster rate. This process is known as **accretion**. The planets and moons in our Solar System were formed in this manner. Our Solar System was formed 4.5 billion years ago.

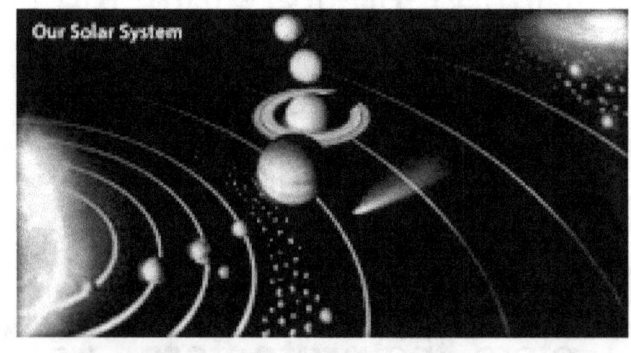

Observations made by Edwin Hubble support the Big Bang theory. In the 1920's, he was able to make observations of the galaxies within our universe with his 100 inch telescope. He was able to show that the galaxies were moving away from each other as well as moving away from other galaxies at accelerating rates. He suggested that the universe was expanding as a result of a Big Bang event that set matter into motion away from its place of origin.

Knowledge and Comprehension
Words to Know

Big Bang Theory:

Matter:

Protostars:

Stars:

Galaxies:

Supernova:

Solar System:

Accretion:

1. Explain how the universe was formed.

2. Which scientist is credited for the discovery of galaxies in our universe? What equipment did he use to do this?

Application, Analysis, Evaluation, and Synthesis

3) Determine and explain the relationship between

a) Stars and Galaxies

b) Stars and Supernova

c) Protostars and Stars

3) Explain how gravity played a role in the formation of galaxies and our Solar System.

4) Find and select evidence within the text to support the Big Bang theory.

	Evidence
1	
2	
3	

5) Could the Big Bang occur again? Explain your answer and use evidence to support it.

Timeline:
Important Events in the Creation of the Life on Earth

ERA	Time Millions of Years Ago	Event
Precambrian Time	4500	The Formation of the Earth and the Solar System
	3600	**The first prokaryote is created**
	2,500-544	**First shelled animals are created**
	544	Mass Extinction
Paleozoic	544-505	Trilobites form
	505-440	**The First Fish formed**
	505-440	Mass Extinction
	440	**Land Plants are formed**
	410-360	**Amphibians are created**
	410-360	Mass Extinction
	360-325	**First Reptiles are created**
Mesozoic Era	248	Mass Extinction
	213-145	**First Birds are created**
	145-65	**First Flowering Plants are formed**
Cenozoic Era	65	Mass Extinction
	65	**The First Mammals Appear**
	60	**The First primates are formed**
	22 million years ago	**The Great Apes appear**

ERA	Time Millions of Years Ago	Event
	5.6 - 4.4	**The first Hominids, Ardipithecus, Evolve**
	3.9	Australopithecus Afarensis (Lucy)
	2.3-1.44	Homo Habilis/Homo Rudolfesis
	1.8 - .1	Homo Erectus
	1.3 - .2	Homo Heidelbergensis
	.5 -.03	Homo Neanderthalis
	.3 to present	**Homo Sapien Sapiens/ Anatomically Modern Humans**

Art Activity: Create a timeline that contains pictures and descriptions of the most important events in the creation of life on Earth. Choose 10 events from the timeline above that you think are the most important. Make sure you indicate how many years ago the events occurred.

Knowledge Building Questions:

1. How many millions of years ago did the Earth form?

2. What happened 5.6 million years ago?

3. When was life first created on Earth? When did this occur?

4. What was the first life form created on Earth?

5. When did single celled organisms transition into multicellular organisms? What were the first multicellular organisms?

6. When did the following animals appear on Earth? Are these animals evolving into simpler or more complex lifeforms? Explain?

    Fish

    Amphibians

    Reptiles

    Birds

    Mammals

7. From which type of animal did primates evolve from? When did this occur?

8. Which type of animal evolved into the Hominids? When did this happen?

9. From which animal group did modern humans evolve from? When did this happen?

10. Did humans evolve early or recently in the timeline of life on Earth?

The History of the Earth

The planet Earth was created about 4.6 billion years ago. Earth, along with the other planets in the solar system, was formed out of the dust, rocks, and gases that were left over from the solar **nebula** that formed our sun and our solar system. Nebulas form from the remains of massive stars that have ended their lives. This massive period of time marks  the development of our planet along with the creation and development of all the lifeforms that are living today and the lifeforms that have gone extinct.

The 4.6 billion year time span has been divided up into smaller **intervals** or periods of time such as eons, eras, epochs, and periods to establish a **geologic time scale** or the time scale of the Earth.

How do we know the age of the Earth? The geologic time scale was established using dating methods that exist today. In 1989, scientists discovered that the oldest regions on Earth are located in continental shields. **Continental shields** are places on the Earth's crust where the rocks have been dated to precambrian times. **Precambrian time** is the oldest period of time the marks the origin or beginning of our planet's existence. Zircon crystals within rocks found in regions in Canada and Australia have been dated back to 4.0 and 4.0 billion years ago. Stony meteorites, meteorites made of stone, found in Antarctica have been dated back to 4.48 and 4.56 billion years ago. **Meteorites** are debris that is left over from comets, or sometimes material that has been ejected into space from impacts on other planets and moons. **Moon rocks**, rocks brought back from the moon have been dated back to 4.6 billion years ago. This evidence helps to confirm the theory that our solar system, along with the sun, the planets, and the moons were formed together, at the same time, 4.6 billion years ago.

Knowledge and Comprehension
Words to Know:

Nebula:

Intervals:

Geologic Time Scale:

Continental Shield:

Precambrian Time:

Meteorites:

Moon Rocks:

1. Describe how the Earth formed.

2. What does the geologic time scale represent?

Application, Analysis, Evaluation and Synthesis

3. Explain where the oldest rocks on earth can be found. What evidence confirms this?

4. How old is our Earth? What evidence supports this? Do you agree or disagree? Explain your thinking.

5. What evidence supports the claim that the Earth, our moon and the solar system was formed at the same time from a solar nebula. Find this evidence in the text. Why is this important?

What is a Fossil?

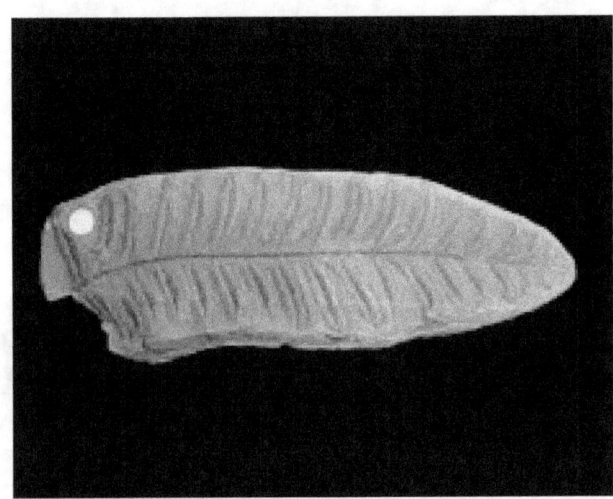

A **fossil** is the preserved remains such as bones or traces of other plants, animals or other organisms from the remote past. The word fossil come from the Latin word "fossilis" which means "obtained by digging." A **trace fossil** is an imprint or a mark that is left by an organism. An example of this is a footprint of a dinosaur or the impression of a leaf. Fossils are important because they allow scientists to study different species of organisms throughout time, how they formed, how they evolved, and the relationships between different types of organisms.

The oldest fossils of organisms have been dated back to 3.48 billion years ago. Scientists, in the 19th century, found that the age of a fossils matched the age of the **rock strata**, or rock

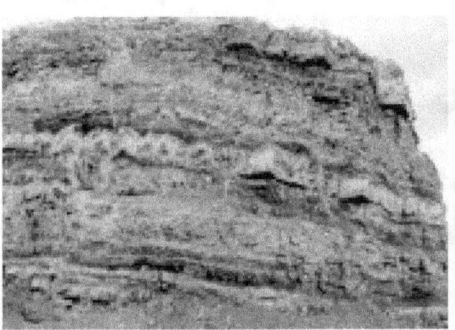

layers, that the fossils were excavated from. This relationship between the age of the fossils and the age of the rock strata led geologists to recognize a **geological timescale,** a timescale based on the study of the Earth's rocks and sediments. They realized that certain rock strata were formed throughout time, and the rock strata could be dated using **radiometric dating** and their absolute age could be determined.

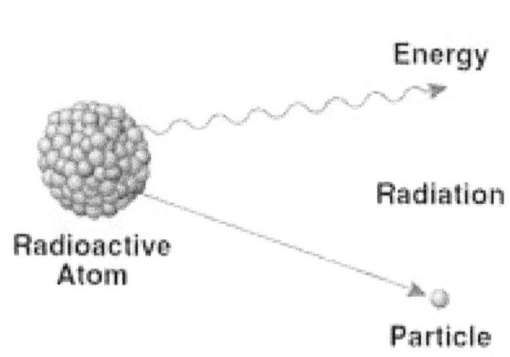

The absolute age of a rock sample can be determined by analyzing the radioactive decay of this material over time. Scientists can estimate the age of the rock if they know what its half-life is. **Half-life** is the time it takes half of a sample of material to breakdown by radioactive decay. **Radioactive decay** is the breakdown of the nucleus within an unstable atom. The unstable atom will lose alpha and beta particles until it becomes stable or has a nucleus that contains the same number of protons and neutrons inside it.

Knowledge and Comprehension:

Fossil:

Trace Fossil:

Rock Strata:

Geological Timescale:

Radiometric Dating:

Half Life:

Radioactive decay:

1. What is a fossil?

2. What is the difference between a fossil and a trace fossil?

3. What does radiometric dating allow a scientist to do?

Application, Analysis, Evaluation, Synthesis

4. Explain what the relationship is between the age of a fossil and the age of the rock or sediment it was excavated from.

5. Why are fossils important to study?

6. If you found fossils in two different rock strata, would you expect them to be the same age? Why or why not?

Precambrian Time: Life on Earth Begins

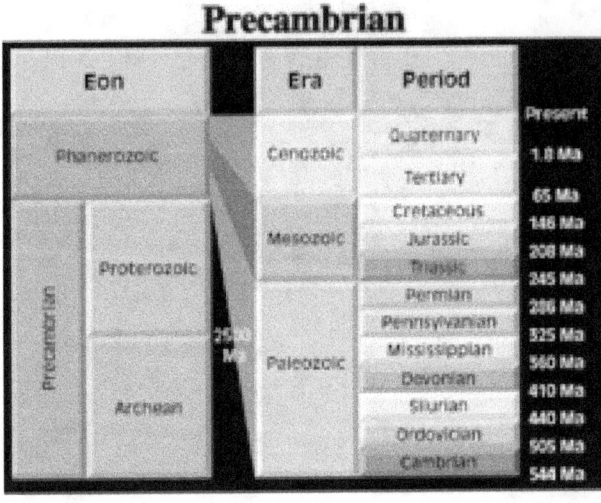

Precambrian time is a period of time that begins with the formation of the Earth from a the remains of a solar nebula 4.5 billion years ago, to 542 million years on the geologic time scale. This span of time is many forms of life were created and their remains of primitive animals and plants were found in the **sedimentary layers** or rock layers. We know these remains as **fossils,** that date back more than 3.42 billion years ago.

The layers of rock are a record that represents the different eras of development and changes that occurred on the planet, as well as, the formation of all its lifeforms. 600 million years ago, the Earth was hot and molten. As the sedimentary layers cooled down on the outside of the planet, a solid outer crust began to form. Water vapor that had been trapped in the atmosphere, slowly condensed, fell to the Earth, and gathered to form the oceans.

Scientists theorize that the first life form was single-celled bacteria that lived within the ocean 3.42 billion years ago. At this time, the environmental conditions on the Earth were volcanic and **anaerobic** or lacked oxygen within the atmosphere and the oceans. It 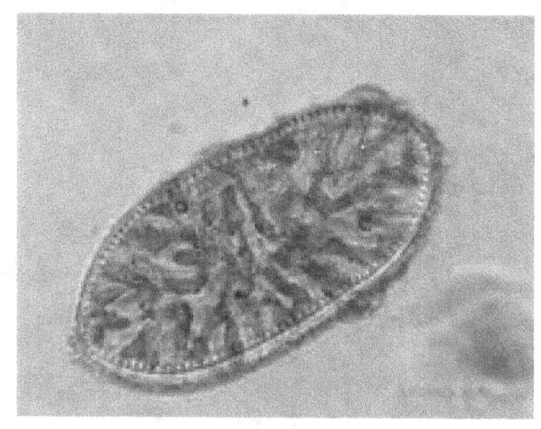 is believed by scientists that all lifeforms on Earth evolved from a primordial, or first, single-celled organism known as an **archeo-bacteria.** These tiny organisms may have used the minerals and gases from volcanic vents for fuel. These vents, a good source of hot mineral rich water and gases, may have served as a catalyst for chemical reactions. A **catalyst** speeds up the rate at which chemical reactions occur. Eventually, formed the archeo-bacteria, the first primordial and simple organisms to inhabit the Earth.

Archeo-bacteria consisted of a **cell membrane**, an outer layer of molecules able to regulate the movement of chemicals in and out of the organism. This outer layer surrounded organic molecules that underwent specific chemical reactions such as the conversion of molecules into useable energy. One of the many chemical reactions may have involved **self-replication** substances or macromolecules that were able to reproduce by themselves. Examples of self-replicating substances within cells are nucleic acids such as DNA and RNA.

Source:

http://www.britannica.com/EBchecked/topic/474302/Precambrian-time

Knowledge and Comprehension
Words to Know:

Precambrian Times:

Sedimentary Layers:

Fossils:

Anaerobic:

Archeo-bacteria:

Catalyst:

Cell Membrane:

Self-replication:

1. Describe what Precambrian time is? What important event occurs during this time?

2. When did the first organisms on Earth appear?

Application, Analysis, Evaluation and Synthesis

3. What evidence do we have of past life that existed on Earth? Where was this evidence found?

4. If the Earth first formed 4.5 billion years ago and the first life form appeared about 3.42 billion years ago, explain why it took 1 billion years for life to be created.

5. Consider the claim: Life on Earth formed on its own then evolved into many lifeforms. Do you agree or disagree? Explain your thinking.

6. Describe the environmental conditions on Earth that allowed the first lifeforms to be created. How could this organism have changed and evolved through time to give rise to organisms that breathe oxygen to survive?

Cyanobacteria:
Making Oxygen on Earth

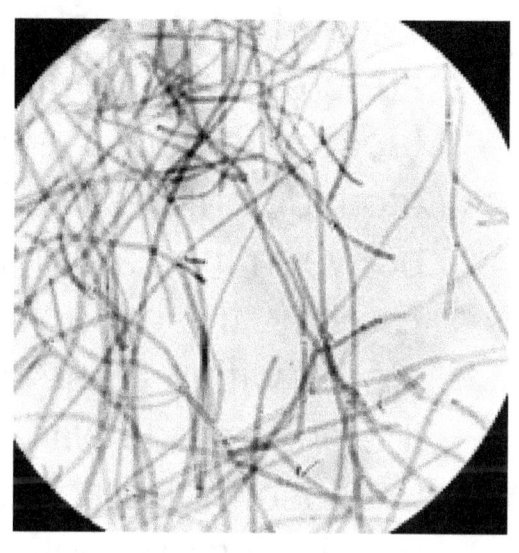

Cyanobacteria, a blue-green bacteria, is thought by scientists to be the first organisms to have introduced oxygen onto the Earth. **Oxygen** atoms contain 6 protons in their nuclei and 6 electrons in their outer shells. They readily share electrons with other atoms such as carbon, nitrogen, and hydrogen in order to fill their outer shells. The early Earth was marked by many volcanic eruptions that filled the atmosphere with the compounds carbon dioxide, hydrogen sulfide gas, ammonia, and nitrogen gas. Cyanobacteria are believed to be the first organisms capable of using **photosynthesis** to produce fuel source.

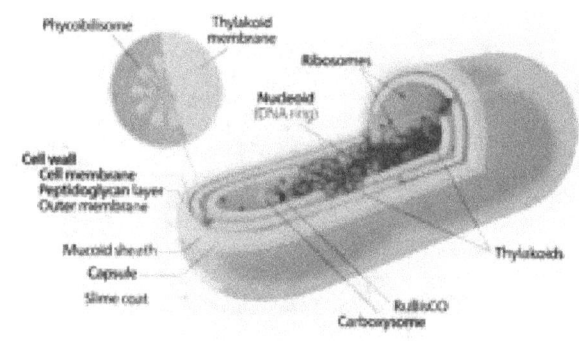

Photosynthesis is the process of taking the compounds carbon dioxide and water and making the sugar glucose.

Glucose is a chain of 6 carbon atoms that have been assembled into a ring structure through bonds. These **bonds** are attractions that are created through the sharing of **electrons** negatively-charged sub atomic particles within the atom, between the carbon atoms.

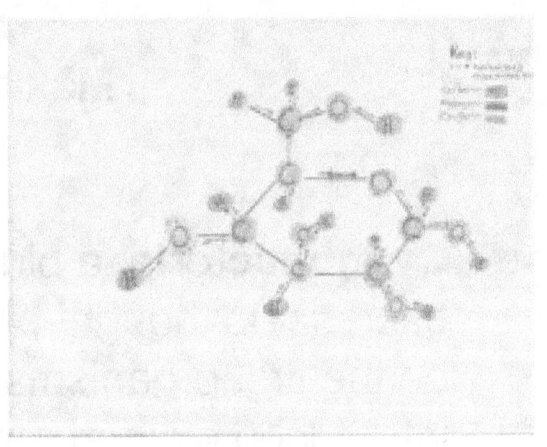

Energy, the capacity to do work, is stored within these bonds and with other bonds such as the ones made with oxygen atoms and hydrogen atoms. Glucose molecules are stored within the organisms when needed as an energy source at a later time. When the fuel source of cyanobacteria becomes low, it breaks down glucose to use as food, it gets its energy from the breaking of each one of its bonds.

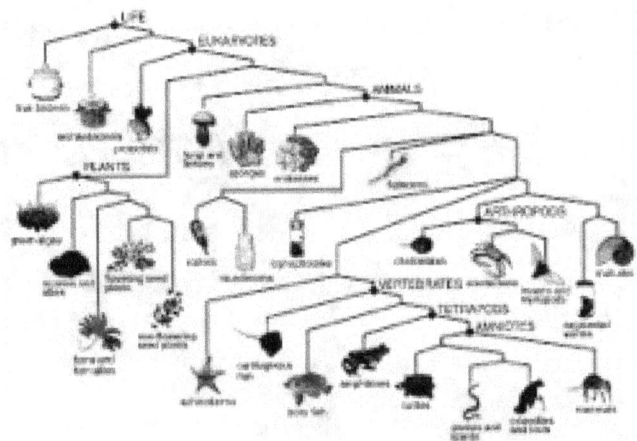

By using photosynthesis and producing oxygen gas as a waste product, cyanobacteria have been believed to convert the early atmosphere, once anaerobic and toxic, into an oxygen-rich atmosphere. Once this transition occurred, it opened the door for the formation of more complex life forms that used oxygen to survive both in the sea and on land. This event stimulated the beginning of **biodiversity**, or many different kinds of organisms, on earth with lifeforms of different compositions, structures, and characteristics. This event also lead to the near extinction of species that could not survive in the presence of oxygen.

Knowledge and Comprehension
Words to Know:

Cyanobacteria:

Oxygen:

Photosynthesis:

Glucose:

Bonds:

Electrons:

Energy:

Biodiversity:

1. What is cyanobacteria?

2. Why were cyanobacteria important to the early Earth?

3. What is photosynthesis?

Application, Analysis, Evaluation and Synthesis

4. Explain how photosynthesis helped to change the the composition of the atmosphere. What new gas was introduced by this process?

5. Explain how glucose "stores" energy. What must be done in order to release this energy?

6. Explain how cyanobacteria benefits from making and storing glucose.

7. Explain the cause of biodiversity on Earth. Why is this important?

The **Proterozoic Eon**

The **Proterozoic Eon** is a division of time of the geological time scale that represents the time just before complex life was created on Earth. It encompasses the middle to end of Precambrian Era.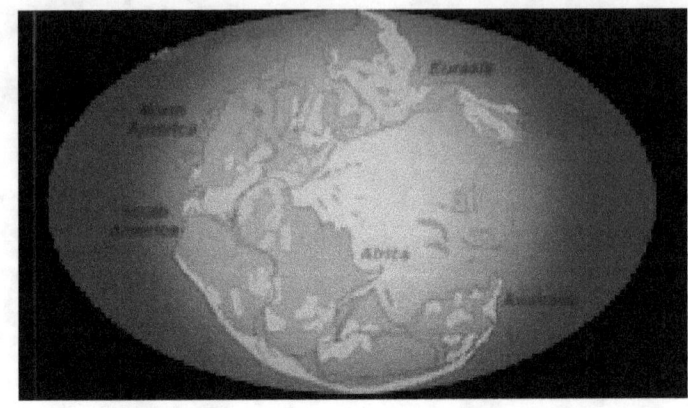
The name **Proterozoic** means " earlier life." It started 2500 million years ago and ended 542 million years ago with the first mass extinction.

The eon is divided into three segments: the Paleoproterozoic, the Mesoproterozoic, and the Neoproterozoic. The Paleoproterozoic is well known for the transition of the Earth to an atmosphere that was oxygenated with the help of cyanobacteria. The Neoproterozoic (635-542 million years ago) is marked by periods of **glaciation** or periods of time when the Earth's climate transitioned from being warm to an ice age climate. **Ice ages** are characterized by freezing temperatures and the development of massive ice sheets that start at the North and South poles and grow towards the equator of the Earth. This time span also includes the hypothesis of the "**Snowball Earth**" or a time in history when the entire Earth was covered with ice sheets and was white as a snowball.

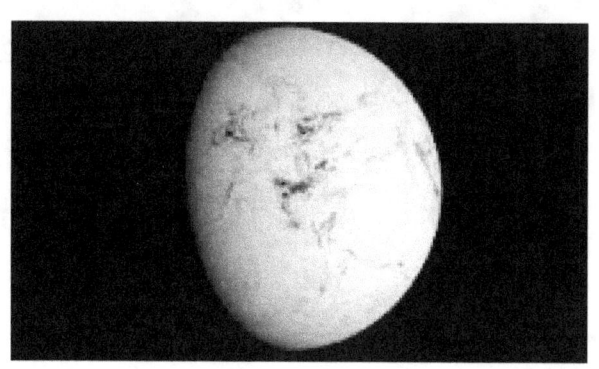

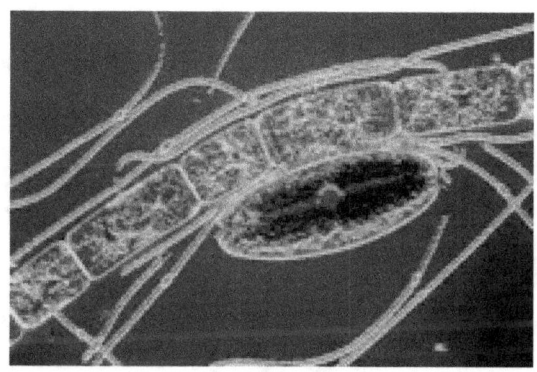

The formation of soft-bodied multicellular organisms occurred during the Neoproterozoic. These organisms were the direct result of the evolution of **single-celled** (one-celled) **organisms** to **multicellular** (many-celled) **organisms**. During this period of time, the first symbiotic relationships between chloroplasts and eukaryotes and plants occurred. This was an important event because it allowed the cells of eukaryotes and plants on land to use cellular respiration to breakdown glucose, a simple sugar, into energy. This process occurs only in the presence of oxygen. This allowed for the evolution of more complex, multicellular lifeforms, such as fungi, plants and animals to form.

Fossils of the first animals found by scientists include trilobites and archeocyathids appeared at the end of the Proterozoic Eon about 542 million years ago.

Trilobites are extinct marine arthropods and are the earliest known group of arthropods. They are invertebrate animals that have a hard exoskeleton or external skeleton. Their skeletons are made up of chitin which is a different form of glucose.

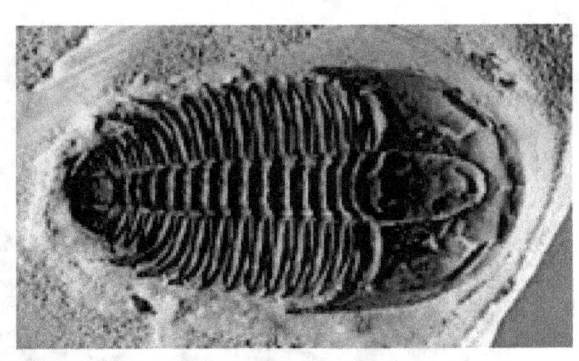

Arthropods formed the phylum Anthropoda which include members such as spiders, insects, and crustaceans such as crabs and lobsters. Crusteceans also contain calcium carbonate that has biomineralized or produced by living organisms. **Archeocyathids** or "ancient cups" in Greek are extinct reef building marine organisms that lived in warm, tropical water.

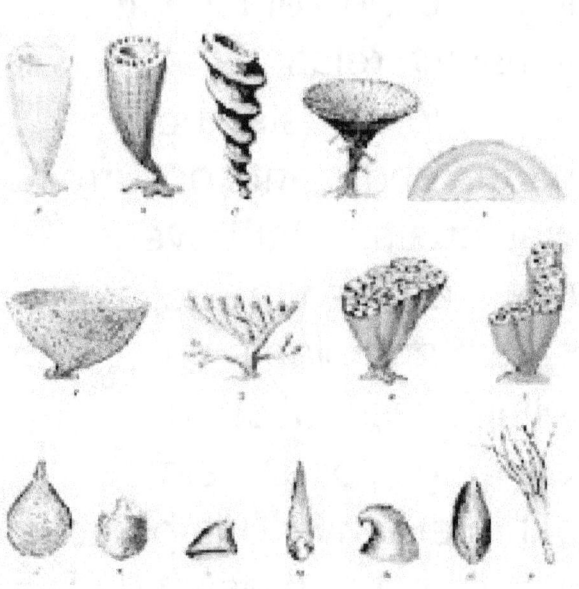

Knowledge and Comprehension:

Proterozoic Eon:

Proterozoic:

Glaciation:

Ice Ages:

Snowball Earth:

Single Celled Organisms:

Multicellular Organisms:

Trilobites:

Archeocyathids:

1. Describe the Proterozoic Eon. What is it? When did it occur. What important events happened during this period of time?

2. Describe what the climate on Earth was like during this period of time.

Application, Analysis, Evaluation, Synthesis

3. Explain how multicellular organisms evolved during this period of time. Why do you think this occurred?

4. Explain the theory of Snowball Earth. Is it possible that this type of climate could happen again on Earth?

5. Describe the new lifeforms that were created during the Proterozoic Era.

6. Predict what the next life form to evolve after the Proterozoic Era would have been. Explain how and where the animals within this eon would have evolved. How would they have through adapted to the environment and the climate?

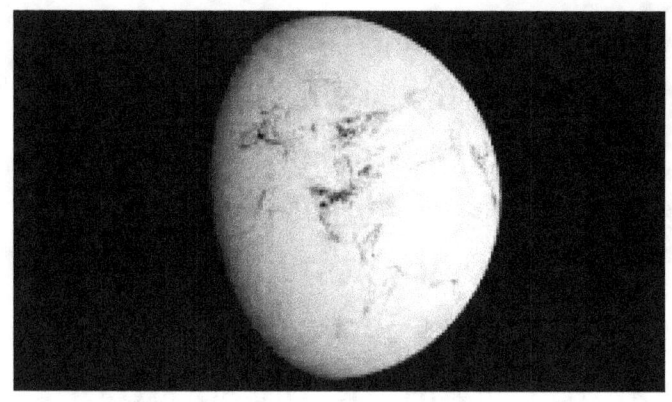
The **Proterozoic Eon** is a division of time of the geological time scale that represents the time just before complex life was created on Earth. It encompasses the middle to end of Precambrian Era. The name **Proterozoic** means " earlier life." It started 2500 million years ago and anded 542 million years ago with the first mass extinction.

The transition of the Earth to an atmosphere that was oxygenated with the help of cyanobacteria was an important event that occurred during this time period. The Neoproterozoic (635-542 million years ago) is marked by periods of **glaciation** or periods of time when the Earth's climate transitioned from being warm to an ice age climate. **Ice ages** are characterized by freezing temperatures and the development of massive ice sheets that start at the North and South poles and grow towards the equator of the Earth. This time span also includes the hypothesis of the "**Snowball Earth**" or a time in history when the entire Earth was covered with ice sheets and was white as a snowball.

The formation of soft-bodied multicellular organisms occurred during the Neoproterozoic. These organisms were the direct result of the evolution of **single-celled** (one-celled) **organisms** to **multicellular** (many-celled) **organisms**. During this period of time, the first symbiotic relationships between chloroplasts and eukaryotes and plants occurred. This was an important event because it allowed the cells of eukaryotes and plants on land to use cellular respiration to breakdown glucose, a simple sugar, into energy. This process occurs only in the presence of oxygen. This allowed for the evolution of more complex, multicellular lifeforms, such as fungi, plants and animals to form.

Knowledge and Comprehension:

Proterozoic Eon:

Proterozoic:

Glaciation:

Ice Ages:

Snowball Earth:

Single Celled Organisms:

Multicellular Organisms:

Trilobites:

Archeocyathids:

1. Describe the Proterozoic Eon. What is it? When did it occur. What important events happened during this period of time?

2. Describe what the climate on Earth was like during this period of time.

Application, Analysis, Evaluation, Synthesis

3. Explain how multicellular organisms evolved during this period of time. Why do you think this occurred?

4. Explain the theory of Snowball Earth. Is it possible that this type of climate could happen again on Earth?

5. Describe the new lifeforms that were created during the Proterozoic Era.

6. Predict what the next life form to evolve after the Proterozoic Era would have been. Explain how and where the animals within this eon would have evolved. How would they have through adapted to the environment and the climate?

Mass Extinction

Mass extinction is defined as the death of many different organisms during a specific period of time. A mass extinction is preceded by an extinction event. An **extinction event** is an event that has the potential to sharply reduce the amount of life that exists worldwide. A mass extinction decreases biodiversity by reducing the number of species that exist on our planet. A **species** is defined as the largest group of organisms that are able to interbreed and have fertile offspring.

There has been a total of 5 mass extinctions during Earth's geologic history. Every major extinction in Earth's geologic history has given rise to new species, a process called **speciation**. It is estimated by scientists that 96% of the species that ever existed on the Earth, have gone **extinct** or are no longer living.

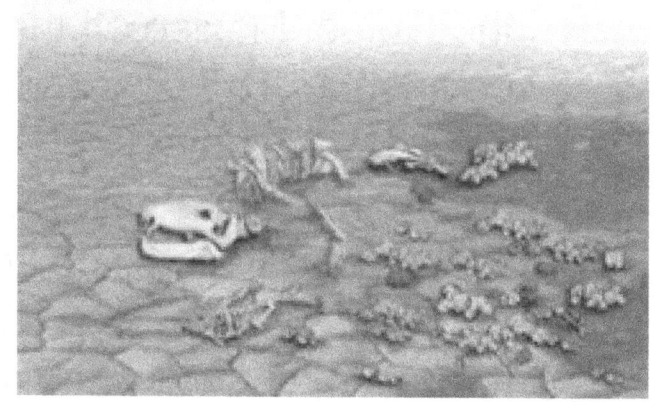

Fossils within the fossil record is evidence of species that had roamed the Earth at one time in the past. The excavation of dinosaur fossils and study of the sediment layers in Colorado revealed a surprising fact. Scientists found that the layer of sediment below where the dinosaurs were dug out containing a thin layer of compacted clay. The clay represents a layer of ash that had fallen before the death of the dinosaurs. This thin layer of clay has also been found in different places around the world. This clay was radiocarbon dated to about 65 million years old.

A different set of scientists discovered that a huge crater in the Gulf of Mexico off the Yucatan peninsula. The crater was estimated to have been created by an asteroid impact about 66 million years ago. The age was determined by radiocarbon dating the rocks that were marked by the impact. Core samples of the sediment within the crater reveal

the presence of the element Iridium which is does not naturally occur on Earth. It is a green layer within sediments that are analyzed. Scientists have also found andecite, a mineral, that had been melted into glass in the sediment on the Yucatan and on the island of Haiti. Melted minerals are consistent with meteor and asteroid impacts.

Physicist Luis Alvarez and his son Walter analyzed these two important pieces of evidence and theorized that these two events were connected: a mass extinction of the dinosaurs may have occurred as a result of the asteroid impact in the Yucatan. This would explain the layer of clay/ash that may have been the settling of the debris that the impact had thrown into the atmosphere and the untimely death of the dinosaurs. They reasoned that the debris that was thrown into the atmosphere blocked the sunlight for several years, cooling the Earth, and reducing the available plant life for the dinosaurs. The dinosaurs, as a result, may have starved to death and may not have been ale to maintain a high enough body temperature to survive.

Sources:

http://en.wikipedia.org/wiki/Extinction

Focus Questions:

1. Describe what a mass extinction is.

2. Identify what the effects of a mass extinction is on the species of plant and animal life on Earth.

3. What are fossils? Why are they important?

4. Explain why the dinosaurs went extinct. Support your answer with evidence from the text.

5. Explain how scientists know that a mass extinction of the dinosaurs may have occurred as a result of the asteroid impact in the Yucatan. Find evidence in the text that supports this conclusion.